IB MIDDLE YEARS PROGRAMME

T0302872

Global Issues

Project Organizer

1

Series Editor
Barclay Lelievre
Mike East

Name:

OXFORD
UNIVERSITY PRESS

OXFORD
UNIVERSITY PRESS

Great Clarendon Street, Oxford OX2 6DP

Oxford University Press is a department of the University of Oxford.
It furthers the University's objective of excellence in research, scholarship,
and education by publishing worldwide in

Oxford   New York

Auckland   Cape Town   Dar es Salaam   Hong Kong   Karachi
Kuala Lumpur   Madrid   Melbourne   Mexico City   Nairobi
New Delhi   Shanghai   Taipei   Toronto

With offices in

Argentina   Austria   Brazil   Chile   Czech Republic   France   Greece
Guatemala   Hungary   Italy   Japan   Poland   Portugal   Singapore
South Korea   Switzerland   Thailand   Turkey   Ukraine   Vietnam

British Library Cataloguing in Publication Data

Data available

ISBN: 9780-19-918079-0

20 19 18 17 16 15 14 13 12 11

Printed in India by Multivista Global Pvt. Ltd

Paper used in the production of this book is a natural, recyclable product made
from wood grown in sustainable forests. The manufacturing process conforms to
the environmental regulations to the country of origin.

Author acknowledgments
Thanks to Michael Etheridge and CIS science gang — champions and innovators of
all things MYP. *Barclay Lelievre*

Thank you to my family and the Colegio Internacional de Caracas for the
unwavering support you have given me. *Mike East*

The publishers would like to thank Talei Kunkel and Lisa Nicholson for their advice,
and the International Baccalaureate and Anita Knight, Isabel Machinandiarena,
Patrick Sweeney, and Annie Termaat for permission to reproduce the learner profile
on page 5 which originally appeared in *MYP Interact* (International Baccalaureate,
2008).

Companion website:
www.OxfordSecondary.co.uk/myp

Contents

Introduction

This is the first of five Project Organizers which focus on interdisciplinary learning. The series reflects key aspects of the philosophy and approach of the IB Middle Years Programme, including: being internationally minded, demonstrating academic honesty, and developing the qualities of the IB learner profile:

Interdisciplinary learning

Most often in school you will be timetabled to study different subjects at different times. In life you will be mixing the skills and knowledge from these different subjects to understand things and solve problems. The projects in this book encourage you to use more than one subject to approach the unit questions. The first page of each unit shows the unit question and the two or more focus subjects.

International-mindedness

Today's students need to explore a blend of the local, the national and the international. We only have one planet and the way we act affects it and all life upon it. The chapters in this book will show the links between us and people and places all over the world.

Academic honesty

We all want others to think highly of us. Academic honesty is a set of values and skills that promote personal integrity when doing exams, assignments and homework. By following these values and skills we demonstrate that we are honest and principled.

The IB learner profile

The International Baccalaureate aims to develop internationally minded people who, recognizing their common humanity and shared guardianship of the planet, help to create a better and more peaceful world. IB learners strive to be:

Inquirers
You are curious and ask important questions to inquire into the world around you. You research independently and love learning throughout life.

Knowledgeable
Through your keen exploration of local and global issues you build an in-depth knowledge and understanding across all subject areas.

Thinkers
You think both critically and creatively to help solve problems and make responsible decisions.

Communicators
You are able to understand and express yourself confidently in more than one language. You work well and enthusiastically in team situations.

Principled
You demonstrate honesty, a sense of fairness and respect towards those around you. You take responsibility for your own actions.

Open-minded
You take pride in who you are. You are respectful of others' opinions, traditions and values. You consider more than one point of view when making decisions.

Caring
You are considerate towards the needs of others. You are committed to making a positive difference to others and to the environments.

Risk-takers
You are confident and show courage in new situations. You are keen to try new things. You defend your own beliefs strongly.

Balanced

You recognize the importance of caring for yourself, balancing your physical, emotional and intellectual self (all parts of you!).

Reflective

You think carefully about how you learn through different experiences. By being able to recognize your strengths and limitations you can set goals for further learning and development.

How to use this book

This book is to help inspire and structure interdisciplinary work on global themes. You will gradually populate and personalize your Project Organizer throughout the year, with the completed organizer acting as a record of your interdisciplinary work, along with the folders and other pieces of work that you build up. We hope that you enjoy the units and the challenges that they present!

1 Eradication of poverty

Unit question
Are you a 'have' or a 'have not'?

Subject focus and objectives

Language A
➔ Can you use language to narrate, describe, explain, argue, persuade, inform, entertain and express feelings?
➔ Can you describe the conflicts or problems in the works you have studied, categorize them and then relate them to a world issue?

Humanities
Global awareness
Can you explore issues facing the international community and recognize issues of equality, justice and responsibility?

Areas of interaction
Health and social education
You are expected to develop an awareness and understanding of yourself in the wider society in the context of poverty.

Approaches to learning
Reflection
Self-awareness and self-evaluation through reflecting at various stages of your investigation

Contrasting housing in São Paulo, Brazil

Starting points

Your planet

⇨ The Earth weighs 200,000,000 million tons and is moving through space at 66,700 miles per hour.

⇨ 70% of the Earth's surface is ocean.

⇨ 6,700,000,000 people live on the 30% of the Earth's surface which is land. This is the world population.

1   How many countries are there in the world? _____

2   Which country do you live in? Write the name in full here in your own language and in one other language:

_____

_____

Development is the process of change for the better, and is (or should be) the aim of governments, charities, and you and me.

Development happens all over the world, in every country.

Every country is at a different stage of development.

The difference in development between the richest and poorest countries is very big.

The internet has a huge and growing amount of information about every country and every topic you will be investigating. So your challenge will not be finding information. It will be:

⇨ deciding which information you are really interested in
⇨ sorting out what is important enough to help you answer your question
⇨ deciding if the information you find is likely to be accurate and reliable.

Many organizations collect information about different countries and put it on their websites. They include:
⇨ the government in the country, or a group of governments
⇨ news organizations
⇨ charities such as Oxfam
⇨ worldwide organizations such as the United Nations, Unesco and the World Bank
⇨ websites such as Google Earth.

DID YOU KNOW?

Eight **Millennium Development Goals (MDGs)** were agreed by the 192 United Nations member countries. They include achieving primary education for everyone, and eradicating extreme poverty and hunger. The aim is to achieve these goals by 2015.

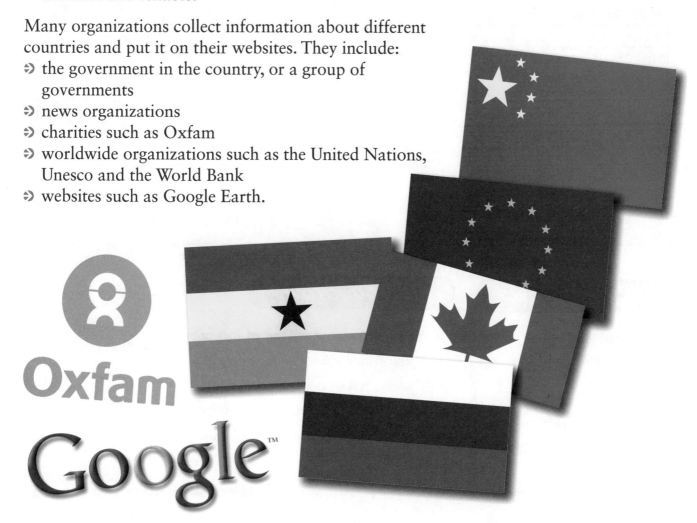

Before beginning to answer the unit question, take a moment to think about it.

---

1 Would you consider yourself to be a 'have' or a 'have not'? Include a brief explanation.

_____

_____

_____

_____

There are many ways we can define the word poor. In a group, use the following space to do some brainstorming. Each member of the group should record a version of the proceedings. You may wish to highlight your contributions.

poverty • noun the state of being extremely poor.

Brainstorm

A	B
What questions would you ask someone to find out if they were poor?	What information would you need to find out to determine if a country is poor?

DID YOU KNOW?

More than one billion people in the world live on less than one US dollar a day. Close to 3 billion struggle to survive on less than two US dollars per day.

DID YOU KNOW?

If the world were a village of 100 people:

• 14 would not have enough to eat

• 31 would have no electricity

• 17 would not have clean safe water to drink

• 20 would be Chinese, 16 would be Indian, and 4.5 would be from the USA.

Reflection

Now that you've had a chance to consider what poverty is, it might be useful to learn directly from students around the world about what kinds of difficulties confront them. You may wish to do some research and uncover some of these stories on your own. Your teacher may also have materials for you to review. Alternatively you could visit www.mylifeisastory.org. This is a website sponsored by UNESCO (United Nations Education Scientific and Cultural Organization).

If you navigate to the Life Stories section, there are childrens' faces you can click on, each with a brief story to tell.

In your research, choose two different children from two countries other than your own. After reading their accounts, make a list in the table below of all the things they mentioned that you think might make their lives difficult or challenging.

> How confident are you about the reliability of information on a UN-sponsored website? What, if anything, does the .org domain extension tell us? What if it had been .net, .com, or .info?

Describe

Story A	Story B

Categorize

Highlight any of the above difficulties that are related in some way to poverty.

Reflection

Take this opportunity to think about your own story, some of the things you feel lucky to 'have' and some of the difficulties you face in your life. Following a similar format to the pieces you have read, write a few paragraphs about your own circumstances.

Before poverty can be eradicated, it must first be identified. When trying to determine how poor an individual or a country is, certain indicators are used. If you visit the United Nations Millennium Development Goals website, you can see the specific ways in which they propose to deal with poverty and some of the indicators they use to determine poverty. The web address is *www.un.org/ millenniumgoals* and there is a link on this page to 'End Poverty and Hunger.'

indicator • noun a thing that indicates a state or level

Reflection

How many of these indicators were similar to those you identified in your brainstorming? Which indicators hadn't you thought of?

Data collection and presentation

1. Decide on a format for a table to display these indicators. You will need to leave space for one of the countries of the students you read about above, or another of your choice, and for the country in which you are currently studying.

2. You should annotate your table by providing definitions or explanations for each indicator. Wherever possible, you should try to include the source of your information.

3. You will need to carry out some research on your own, or with guidance from your teacher or librarian to try to fill out as much of the table as possible. Keep in mind that as the author of these goals, the UN has statistics on most countries.

The UN building in New York city

Analysis

Use the data you have collected to determine if your country and the other you selected is a 'have' or a 'have not'. Include an explanation. This might require you to prioritize. This means make a judgement on which indicator is the most or least important in determining poverty.

Extension

You could do some further research to discover whether or not poverty exists even in countries that are considered 'haves'. Does living in a 'have' country guarantee you will not experience poverty?

Some countries have far more people in them than others, so the indicator of GDP for each person (or GDP per capita) is a fairer indicator to compare.

Just because a country is wealthy doesn't mean that the wealth is spread evenly between its people.

> **GDP, or gross domestic product** • noun The total value of goods and services a country produces in a year. Goods are products such as toys, or naturally occurring things like oil or gold. Services are things like tourism, or doing people's accounts for them.

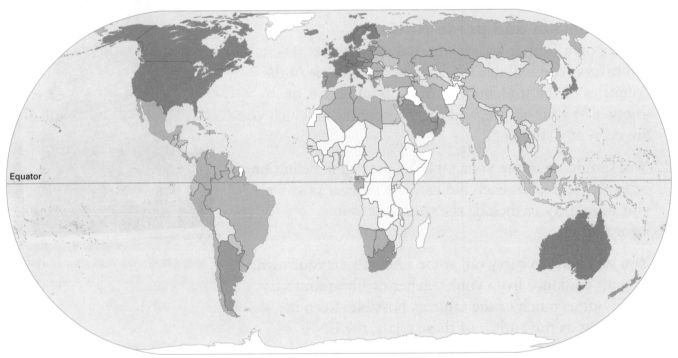

GDP is often used as a measure of the wealth of a country

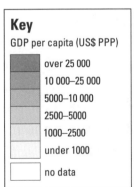

Key

GDP per capita (US$ PPP)

- over 25 000
- 10 000–25 000
- 5000–10 000
- 2500–5000
- 1000–2500
- under 1000
- no data

Our project plan

Please take a moment to review your initial response to the guiding question, your brainstorming results, the readings you completed about students in other countries, and the paragraphs about your own circumstances that you have recently completed.

Your project is to create a piece of writing or other presentation summarising your, or your group's, response to the unit question:

Are you a 'have' or a 'have not'?

The following question might help your planning.
What makes someone a 'have' and someone else a 'have not'?
You may wish to address some or most of the following points as you formulate your response.

- How do you define poverty?
- Did any of the stories you read stir any feelings in you?
- What did the students in the stories you reviewed have in common?
- How were their stories different from each other?
- How were their stories similar or different from your own circumstances?
- What, if anything, did you find surprising or troubling about their accounts?

Extension

You could do some research on the topic of 'ethics'.
If you feel you are a 'have' do you have an ethical responsibility to do something for the 'have nots'?

If you are a 'have not', do you think the 'haves' are ethically responsible for helping to change your circumstances?

When you have completed your project work use the questions on pages 16 and 17 to evaluate how it went and what you might do differently next time.

Project evaluation

1 What did I enjoy about this unit, and why?

_____

_____

_____

_____

_____

_____

2 What aspect didn't I enjoy and why?

_____

_____

_____

_____

_____

_____

3 What did I/what did our team do really well?

_____

_____

_____

_____

_____

_____

4 What did I/what did our team need to improve?

_____

_____

_____

_____

_____

_____

_____

5   What would I do differently next time?

_____

_____

_____

_____

_____

_____

6   What challenges did I/our team face while working on this unit?

_____

_____

_____

_____

_____

_____

_____

7   What do I know now, that I didn't know before working on this unit?

_____

_____

_____

_____

_____

_____

_____

8   Why does this topic matter?

_____

_____

_____

_____

_____

_____

_____

The two faces of poverty – 'have nots' in 'have' countries

Living in a 'have' country does not guarantee that you will not experience poverty. There are over 350 million indigenous people –5% of the world's population – many of whom live in developed countries, yet they account for 15% of those considered extremely poor.

Native people historically lived a modest but self-sufficient existence. They have since often been displaced and decimated by colonialism and exploitation. Their situation has been made worse through continued racism, and failed government policies.

If the United Nations Millennium Development Goal of eliminating extreme poverty is to be met, then everyone needs to be made aware of those most affected. However, there is no specific mention of indigenous people in this MDG. Your job will be to put a face to these 'have nots' in 'have' countries.

indigenous • adjective originating or occurring naturally in a particular place; native

A Baka family in Cameroon

Choose one from the list below:
Canada   United States   Australia   Japan   New Zealand   Mexico   China   Taiwan   Greenland (Denmark)   Brazil   Bolivia   Cameroon

Do some research on their indigenous populations and prepare your report. This should include the following:

1 **Pre-colonial history**

2 **Time of contact with colonists**

3 **Post-colonial history**

4 **Economic status indicators – GDP, unemployment, etc.**

5 **State-sponsored stimulus programs**

6 **Indigenous/aboriginal rights**

7 **Ecological issues**

8 **Future outlook**

2 Refugees and migrants

Unit question
How should I welcome people into my community?

Subject focus and objectives

Language B:
Writing
Can you
⇨ practice writing texts that fulfill particular requirements.

Music
Knowledge and understanding
⇨ Do you understand and describe concepts and processes that support your work?

Application
⇨ Can you use ideas and artistic conventions to create, perform and/or present art?

Physical Education
Use of knowledge
Can you use your knowledge to identify and assess the impact of factors that influence situations, and solve simple problems in familiar situations?

Language A
Organization
⇨ Can you organize ideas and arguments in a coherent and logical manner?

Area of interaction
Health and social education
You will be forming your own judgements about the problems facing refugees and immigrants. What can we do and how can we get involved?

Approaches to learning
This unit focuses on:
Collaboration
⇨ working in groups—demonstrate teamwork

Why leave home?

The following sentences are to help us understand the lives of millions of people. Reflect on the questions as you read then consider the point at the end.

- After the natural disaster destroyed your home would you cross a river full of crocodiles every day to bring drinking water to your family, or would you go somewhere else to live?
- If every day you were insulted and bullied for being just how you are, would you stay, or look for a place to live where people liked you for being you?
- After soldiers started killing the people in your village would you continue to live there?
- One of your parents has said something and the government told you that they will be arrested and put in prison. Do you leave?

If you choose to go, you would become a refugee.

And so you go.

You and your family leave everything familiar behind and find your life full of uncertainty and worry. You no longer know the people around you; they look different and speak with a strange accent. The countryside around you starts to change too. The uncertainties grow:

- Where can you sleep?
- How can you find food?
- When is it safe to stop traveling?
- Where can you find somewhere to settle down?
- Will you see a friendly face again?

You finally find help: a United Nations official, a charity worker, the police, and a government official, all waiting on the border of another country.

What is it reasonable to ask for? What are you going to ask for? Discuss this in small groups and report back to the class.

Just to be clear about this...

refugee • noun - someone who flees their home because of a fear of being harmed if they stay.

asylum seeker • noun - a refugee applying to live in a new country.

migrant • noun - someone who moves to another country.

grant • verb - to give the right to something, to allow.

Why you probably will not get what you ask for

Activity

When responding to someone's concern it is always good to have evidence. Match the comments above with the statements below.

What do these famous people have in common?

Marc Chagall

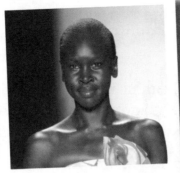
Alek Wek

Wyclef Jean

Isabel Allende

K'naan

Anne Frank (a German Jew who moved to the Netherlands)

K'Naan (a hip hop artist, originally from Somalia who lives in Canada)

MIA (a hip hop artist, born in the UK. Her parents are Tamil refugees from Sri Lanka.)

Maria von Trapp (of the Trapp Family Singers, and *The Sound of Music* fame who left Austria and lived in the USA)

Albert Einstein (a scientist, a German Jew who lived in the USA)

Piet Mondrian (a Dutch painter who moved to New York during World War Two)

Marc Chagall (a Russian Jewish painter who moved to France)

Alek Wek (a supermodel from Sudan who migrated to the UK and then the USA)

Wyclef Jean (a singer from Haiti, of the music group the Fugees—short for refugees—who lives in the USA)

Sitting Bull (Native American chief who took refuge in Canada)

Christopher Wreh (from Liberia, he played for the English football club Arsenal)

Isabel Allende (Chilean writer of *The House of the Spirits* who moved from Peru)

Rigoberta Menchú (a Guatemalan who fled to Mexico and who won the Nobel Peace Prize in 1992)

They were all refugees.

Choose one or two of these people and investigate their lives. Consider what they have added to the world and create a list of their achievements in your folder.

Our project plan

The following activities focus on language, music and sport and together provide background to help you approach the unit question:

How should I welcome people into my community?

Activity 1

A number of refugees arrive in your city. They have been granted asylum (see the definitions on page 20), and begin the long process of integrating into their new community—your community.

Through the community and service program your school gets involved and seeks your help. The refugees have limited language skills and need to be taught the basics of life in your country.

Your job is to produce an easy-to-read and visually attractive leaflet that will explain how to complete a task successfully. These tasks could include:

- How to post a letter
- How to open a bank account
- How to use a coin-operated public laundry
- How to open and use an email account
- How to use the internet
- How to use the bus service
- How to use the trains/metro
- Leisure activities in your area

Buying a metro ticket

The task

Brainstorm possible tasks, make a list and take one each.

Discuss what makes a good piece of writing and study the typical features of leaflets: the vocabulary, the grammar/syntax, the register and the layout.

After you have produced your leaflet you should present and explain it to other members of your class. They could role play the former refugees and ask questions to the presenter.

Activity 2

When refugees and migrants move to a new country they bring their culture with them, of course. When cultures mix interesting things usually happen. This activity will look at the way new music comes from this mix of cultures.

For example: on the Caribbean island of Cuba, the Spanish colonialists of the 1700s had brought the music and dance style that was popular in Europe. Later in the 1850s there was a wave of migration from nearby Haiti to Cuba. The Haitians introduced a greater African influence, through their instruments and their interpretation, to the colonist's music.

Little by little, this blending of two (and more) influences evolved into what today is a worldwide phenomenon: salsa.

The task

➡ Investigate the rhythm and scales of two or more types of music that interest you from different cultural traditions. Present these to your classmates.

➡ Using your voices and your bodies, and/or instruments, and/or composition programs; blend two culturally distinct pieces of music together.

➡ **Formative assessment:** present to the rest of the group, listen to the comments of others and give your comments when it is your turn.

➡ Make changes based on the comments you have received.

➡ Write up your reflections on your presentation and the feedback you received separately and summarize them here.

Salsa dancers

Salsa band

Remember, our unit question is:

How should I welcome people into my community?

A world without salsa, or any of the existing musical styles we can now enjoy, would be a culturally poorer place.

Activity 3

If migrant culture can create new, exciting forms of music, can the same happen with sport?

Lacrosse is a stick and ball game that originated amongst the Native American tribes of what is now Canada and the USA. It was originally played between dozens or even hundreds of participants. The goals could even be a few miles apart and the game would last from dawn until dusk. The ball was scooped up and carried or hit.

Migrants from Europe quickly became interested in lacrosse and by the 1740s many were playing it. It was these people who wrote a set of 'official' rules in 1867. These rules reduced the playing time and the team sizes. They also standardized the size of the stick and banned sticks lined with lead that had been used to bash the heads of the opposing team.

With a blend of two cultures a new game emerged.

An Indian Ball-Play by George Catlin circa 1846-1850

Anyone who has seen a game of Australian rules football knows what an exciting sport it is to watch. Historians are unsure of how it began, but one theory is that it was brought to Australia by Irish migrants and comes from another game called Gaelic football. Another theory is that it is a blend of ideas from the indigenous Aborigines and the migrants.

Australian rules football

The task

Design a new game that is a blend of rules from two or more other games.

⇒ Look at the resources available to you in your physical education department.

⇒ In groups of three, design a new sport that is team-based.

⇒ Recognize the sources you have used in your work.

⇒ The rules of your new game need to be clear and easily refereed.

⇒ Teach your classmates the game.

⇒ One of you should referee, the other two should coach one of the teams each, including supporting and encouraging all members of that team throughout the activity.

⇒ After playing it with your classmates, listen to their feedback (formative assessment) then, based on their advice, make any necessary changes.

⇒ Play it a second time for the summative assessment.

It is now time to conclude by considering the unit question from different points of view. You will be researching, drafting and then writing a persuasive essay on the unit question:

How should I welcome people into my community?
Discuss the question from different points of view.

We have been looking at ways refugees and migrants enrich the communities they join. There are also arguments against accepting refugees and migrants. It is your turn to make a decision about this.

To begin with complete the following planner.

An eye-catching 'hook' or my interesting start would be:

_____

_____

My thesis, belief statement or the point I will argue for will be:

_____

_____

Arguments for:

_____

_____

_____

_____

_____

_____

_____

_____

_____

_____

_____

_____

Arguments against:

_____
_____
_____
_____
_____
_____
_____
_____
_____

An idea I want to have in my conclusion is:

_____
_____
_____
_____
_____
_____
_____
_____
_____
_____

Research the quotations from experts, facts and statistics you will use to justify your argument.

Now write your essay on separate paper. Don't forget to acknowledge the sources of information that you include.

Here are the criteria that you can use to judge your work:

Content (receptive and productive)
How strong are the arguments you put forward. Do you represent the opinions of both sides?

Organization
Do you follow an essay format? Do you include your sources of information?

Style and language mechanics
Do you write in a formal style? Are you technically accurate?

Project evaluation

What activities I enjoyed in this unit, and why?

_____

_____

_____

_____

What aspect didn't I enjoy so much and why?

_____

_____

_____

_____

What did I/our team do really well?

_____

_____

_____

_____

What did I/our team do less well?

_____

_____

_____

_____

What would I do differently next time?

_____

_____

_____

_____

Which group's project work did I like best and why?

_____

_____

_____

_____

Moving on

The wider issue

This unit question has focused on how much we should help refugees that wish to start a new life in our community.

However, there is **the much bigger part of the story** that we do not have time to tell here. Look at this picture.

Refugee camp in northern Sudan

Perhaps 1% of refugees get the chance of a new life in a new country. Most spend their time in refugee camps in a neighboring country to their own.

They wait for things to get better so they can return home...or wait for the possibility of integrating into the community outside of their camp.

Find out about

- ⇢ conditions in refugee camps
- ⇢ what people's daily lives are like
- ⇢ how many people are refugees and in what countries they are located.

Choose a group of refugees to study and research the reasons why they do not want to or cannot return to their homes.

Useful websites

There is one called the **Global Gang**. It is good if you are working in your second language as the vocabulary is accessible and the graphics and interface allow you to easily navigate and explore the issues. You can read the stories of child refugees and join to get email updates.

Also look at:

- ⇢ The United Nations High Commission for Refugees
- ⇢ Save the Children
- ⇢ One World (www.oneworld.net)
- ⇢ The Refugee Council (a British-based group)

3 Education for all

Subject focus and objectives

Mathematics
Knowledge and understanding
Statistics and probability
Can you find the means and ranges of reported quantities from different countries and explore relationships between them?

Investigating patterns
Can you identify variables, organize data, and describe simple patterns and mathematical relationships?

Humanities
Concepts – Change and global awareness
↪ Can you identify links between causes, processes, and consequences?

↪ Can you explore and recognize basic issues facing the international community including equality and responsibility?

Area of interaction
Health and social education
You are expected to develop an awareness of yourself in a wider society.

Approaches to learning
Information literacy
↪ Can you select and organize information, identify points of view, bias, and weaknesses?

↪ Can you make connections between a variety of resources?

A school in Uganda

The United Nations has set the Millennium Development Goal that by the year 2015 every child can complete a full course of primary schooling.

Most countries in the developing world have shown improvement in this area. In fact the number of children not attending school has dropped from 103 million in 1999 to 73 million in 2006, but there is still work to be done.

The World Bank keeps track of the progress of countries toward this goal and gives them a rating based on how much they have improved, and how far they have to go.

The ratings are as follows:

Seriously off-track These countries are on a path to meeting less than 50% of the target.

Off-track These countries are on a path to meeting more than half of the target but less than 100%.

On-track These countries are on a path to meet the target by 2015 or are currently between 85% and 94% complete.

Accomplished – These countries have already met the target or are more than 95% complete.

Progress of developing countries towards the goal of universal primary education

Problem solve

If you know that there are 154 developing countries, can you determine the number of countries there are in each category? Write your answers next to the appropriate piece of pie.

The majority of the children not enrolled in school are from Sub-Saharan Africa (52%) and South Asia (25%), but no region is immune to this problem.

To begin to answer the unit question we will look at a sample of countries from every geographic region and every stage of completion of the MDG. The two main indicators for universal primary education are i) enrolment rates and ii) completion rates. These represent i) the percentage of children of an eligible age that are enrolled in primary school, and ii) the percentage that manage to complete primary school. See the table on the next page.

Progress on education	Country	Primary enrolment (%)	Primary completion rate
Seriously off-track	Haiti	22%	27%
	Burkina Faso	45%	31%
	Iraq	71%	66%
	Afghanistan	34%	32%
Off-track	Nicaragua	86%	54%
	Congo	55%	57%
	UAE	71%	76%
	Pakistan	68%	63%
On-track	Brazil	93%	76%
	South Africa	87%	82%
	Croatia	87%	90%
	Bangladesh	93%	77%
Achieved	Argentina	99%	100.50%
	Algeria	97%	96%
	Poland	97%	100.30%
	South Korea	99%	104%

Region:

Americas	Africa
Europe/Middle East	Asia

Reflection

Before you begin to analyze this data, it might be helpful to think about why some countries are behind and others are further ahead when it comes to education. Perhaps some of the other issues you've studied in other parts of this book might be a factor? Brainstorm a list of reasons for differences, then get together with a small group and share your answers. Complete both sides of the table.

Stop and think

By studying the data from this sample of countries, in each category you might discover some patterns.

i   Is it fair to compare countries from different regions?

ii Would you feel confident making conclusions based on this small sample?

Brainstorm

Factors that prevent reaching the target (Your ideas)	Factors that prevent reaching the target (Group ideas)

Most collected data is processed in some way in order to search for patterns. One of the most common ways to analyse data is to find an average. There are three different measures of average:

The **mode** is the value in the data that occurs most often.	The **median** is the value in the middle of the data. (You need to arrange the data in order first.)	The **mean** is the sum of all the values divided by the number of values. This is what most people mean by average!

Also important in averages is the **range**. This describes how spread out the data is.

range = highest value - lowest value

From the education data on page 33 find the mean and range for primary enrolment and completion rate in each category. Report these in the table below. Another valuable thing to consider is the spread of the data or its variation. The first one has been calculated here:

Primary enrolment: Seriously off-track values (%)

22 + 45 + 71 + 34 = 172
Number of values = 4
Mean = 172 / 4 = 43
Range = 71 − 22 = 49

Save your full calculations so your teacher can check your work.

Progress	Average primary enrolment (%)	Range of enrolment (%)	Average primary completion (%)	Range of completion (%)
Seriously off-track	43	49		
Off-track				
On-track				
Achieved				

Consider these questions and answer, providing details, on separate paper.

1 Do you notice a pattern in average primary enrolment?

2 Does the size of the range of primary enrolment show a pattern?

3 Do you notice a pattern in average primary completion?

4 Is there a pattern in the range of primary completion?

5 Do the averages appear to be correlated?

A word on statistics

'Statistics can be made to prove anything, even the truth.' *Author unknown*

'Statistics are human beings with the tears wiped off.' *Paul Brodeur*

Statistics help us search for patterns and identify areas where improvements can be made. The two quotes above speak to two of the common problems with statistics and their use as evidence. Firstly, how they can be twisted, and secondly, how they can sometimes be dehumanizing.

Here are a few things you should keep in mind when using statistics.

1 Where did the information come from?

If you were interested in finding out about the dangers of smoking, would you rather get your statistics from a study funded by your national medical association or the national tobacco growers association? Bias and subjectivity are part of every study.

2 What are the statistics actually measuring?

If you are not sure what a statistic represents you should be careful using it. If you came across a statistic stating the mean income in the US is $66,500 you might be tempted to assume that this is the "average" person's wage. The number of extraordinarily rich people in the population tends to inflate or skew the mean. The median income is actually $48,200 which might be a more fitting way to think about the "average" American wage.

3 Does a relationship between two statistics prove cause and effect?

The facing graph appears to show a strong positive relationship between the variables. As *x* increases, *y* increases. Does this mean that *y* increases because of *x*? We must always be careful that there isn't some other variable that is actually at work.

For example, sales of ice-cream and murder rates were found to have a high correlation. Does this mean ice-cream causes people to commit murder? Further research reveals that both ice cream sales and murder rates go up as the seasonal temperature increases. Heat is the true cause for

correlation • 1 a mutual relationship. 3 interdependence of variable quantities

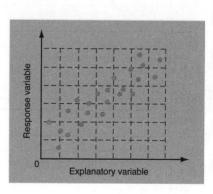

both increases. Ice-cream and murder rates are related only through coincidence.

4 What do statistics really represent?

Many of the statistics we look at in this book describe human deprivation and suffering. It's easy to forget this when looking at numbers and graphs.

Present your data

Graphs– including pie graphs, line graphs, bar graphs and histograms are important ways to display data and help with pattern recognition and data analysis. On the right is an example of a bar graph (used because the data is in categories) showing how birth rate differs by group. A pattern instantly reveals itself. The closer a group of countries are to their goal of universal education, the smaller the birth rate.

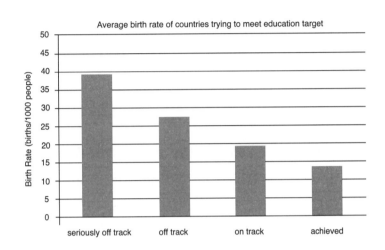

Using the axes below, or using Excel, create graphs to display mean primary enrolment and mean primary completion. Label the axes and include a title.

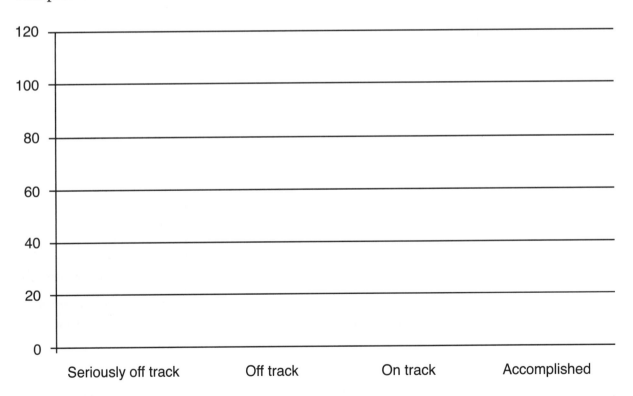

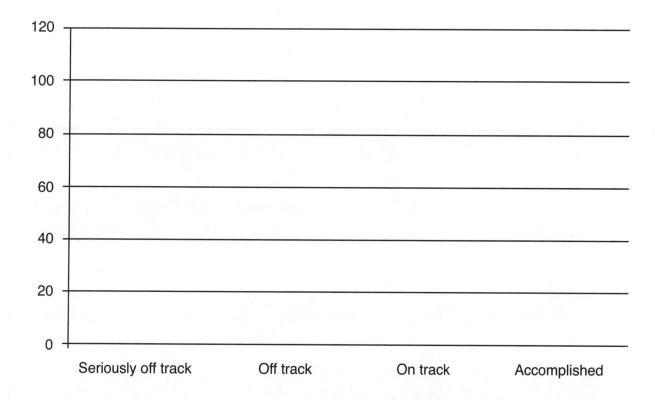

Revisit the questions from page 34 regarding patterns and apply them to the graphs you just completed. Have any of your answers changed? If so, provide details.

In trying to answer the unit question, *'Why doesn't everyone go to school?'* we need to look for factors that might help explain the problem. On page 38 is a table much like the one on page 33 but including a sampling of other human development statistics. These include:

⇨ **Literacy rate** – the percentage of citizens over 15 years of age that can read

⇨ **GDP per capita** – the GDP divided by the number of citizens in the population

⇨ **Birth rate** – the number of live births for every 1000 citizens per year

⇨ **Life expectancy** – the average age at death

⇨ **Population below the poverty line** – the percentage of population living below poverty line

Stop and think

Q: How can completion rates be higher than 100%?

A:Some students repeat grades!

Q: How can countries be considered farther "off-track" if their rates of enrolment and completion are higher?

A: Remember that these are progress reports. So even if some countries numbers are higher, they may not be progressing fast enough, or may even be moving backwards

Progress on education	Country	Enrolment (Primary) %	Primary completion rate	Literacy rate	GDP per capita US$	Birth rate births/1000	Life expectancy (years)	Population below poverty line (%)
Seriously off-track	Haiti	22%	27%	61.0%	$1,109	35.7	57.6	80%
	Burkina Faso	45%	31%	26.0%	$1,084	44.7	51.7	46%
	Iraq	71%	66%	74.0%	$3,700	30.7	58.3	NA
	Afghanistan	34%	32%	28.0%	$1,000	45.8	43.2	53%
Off-track	Nicaragua	86%	54%	80.4%	$2,441	23.7	72.3	48%
	Congo	55%	57%	86.0%	$3,550	41.8	54.5	70%
	UAE	71%	76%	89.0%	$37,000	16	75.9	19.5%
	Pakistan	68%	63%	64.9%	$2,361	28.4	64.3	32.6%
On-track	Brazil	93%	76%	89.6%	$8,949	18.7	71.7	21.5%
	South Africa	87%	82%	87.6%	$9,087	20.23	50.1	50%
	Croatia	87%	90%	98.6%	$14,309	9.6	75.5	11%
	Bangladesh	93%	77%	52.5%	$1,155	28.8	63.5	36%
Achieved	Argentina	99%	100.50%	97.6%	$13,100	18.1	76.4	23.4%
	Algeria	97%	96%	72.0%	$7,426	17	74.6	22.6%
	Poland	97%	100.30%	99.3%	$14,675	10	75.3	17%
	South Korea	99%	104%	98.0%	$22,985	9	78.2	15%

Choose three statistics from the table above that you think might explain why some countries have met their goal and why some are off target. Include an explanation for each of your choices.

Factor 1:	Mean:	Range:
Explanation:		
Factor 2:	Mean:	Range:
Explanation:		
Factor 3:	Mean:	Range:
Explanation:		

You have already calculated the means and ranges for completion and enrolment rates. Correlations can be shown with a scatter plot. The 'line of best fit' on this plot is drawn to show any pattern. The example at the top of page 39 shows primary enrolment rates and paved roads. This might seem like a strange combination, but getting to school is often one of the things that prevent children from attending school.

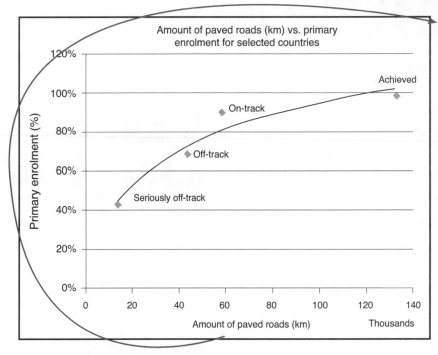

Notice that the "new" factor – roads, is on the x-axis. This is because we want to see if it behaves as an independent variable – the variable that causes a change. If the correlation is strong, we could think about paving more roads to see if it led to better enrolment in schools.

Notice that many of the points are very far from the line. This means that there appears to be a weak relationship or correlation between these variables.

Choose **one** of the factors from your table on page 38 and use the axes below or Excel to create a graph similar to the one above.

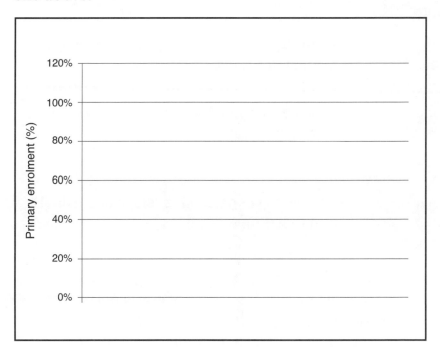

Is there a correlation between the variables you graphed? Provide details.

_____

_____

Our project plan

Did the graph that you created on page 39 help to answer the unit question?

Why doesn't everyone go to school?

Before you answer, you might consider what recommendations the UN has made to solve the problem. Below is a portion of a list entitled *What Needs To Be Done* published in their millennium development goals report.

- increase spending on education to 15 to 20 per cent of national budgets
- eliminate school fees, especially for low income families
- make education a part of humanitarian responses (post-conflict and emergency situations)
- provide children with transportation to and from school when needed
- offer free meals and basic health services at school
- train more teachers and spend more to retain those in the profession
- provide $11 billion US in aid until 2015

So why doesn't everyone – especially girls–go to school? The story below, told by 12 year-old Amina from a small village in Tanzania, hints at several common reasons.

'I wake up very early,' she says. 'I have no notion of time. I sweep the compound, wash last night's dishes. Then I go to the well, but the natural wells are all dry now and we have to walk very far to the artificial wells. We have to wait in a long queue as there are many people. Then I go to the farm to dig or pick cashew nuts. I prepare the day's relish and make ugali for the family. I sometimes get a few hours to play with my friends in the afternoon. I pound cassava or maize for the evening and next day's meal; I then cook supper. After meals I play or listen to adult conversations, especially when there's moonlight. I go to bed when adults go to sleep. Maybe I could go to school, but it is expensive, and my mother will be alone to do all the work.'

You will need to do some research to discover some of the common roadblocks to girls going to school. You might wish to visit some of the websites of some of the following organizations, or your teacher may have materials to share with you.

- Unicef – GAP Report
- Save the Children
- World Bank
- WorldEd.org
- CARE

Once these roadblocks have been identified, you should outline a plan for dealing with the problem of girls not attending school. You will formalize this plan and write a 300-400 word op/ed piece for your local newspaper.

Op/ed pieces are short for opinion/editorial. These kinds of columns appear in newspapers around the world and are written by great thinkers, academics, experts, people in positions of influence, or those whose voice represents a group of people in the population. They are different from newspaper articles, which report the facts only, because they allow the author to express his or her own opinion. By doing so, these pieces can often shape, influence or change the opinions of others, create a meaningful dialogue, and ultimately spur people and governments on to action.

You can find op/ed pieces in every major world newspaper. Do some research by reading a few, and you will soon get a sense of their tone and their power.

Which came first, the chicken or the egg?

This saying is often used to describe situations where it is impossible to say which of the variables is the cause, and which is the effect. For instance does a poor education lead to a low GDP or does a low GDP lead to poor education? Often there is a downward spiral effect, but the answer may never reveal itself.

The graph shows the numbers of girls attending school as a percentage of boys attending. Although there has been improvement, girls continue to be denied schooling. (Why do you think girls are denied this basic right?) It could also be considered a matter of life and death.

The World Bank estimates that for every year of education a girl receives, under-five mortality rates (the number of babies under 5 who die) are reduced by up to 10%! A single year of schooling could potentially save two million of the children these girls will have one day!

Why the dramatic impact? Educated girls are more likely to understand about health, nutrition, and sanitation. They are also more likely to delay marriage, have fewer children, and perhaps most importantly, pass their education to the next generation.

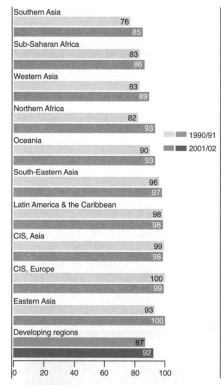

Region	1990/91	2001/02
Southern Asia	76	85
Sub-Saharan Africa	83	86
Western Asia	83	89
Northern Africa	82	93
Oceania	90	93
South-Eastern Asia	96	97
Latin America & the Caribbean	98	98
CIS, Asia	99	98
CIS, Europe	100	99
Eastern Asia	93	100
Developing regions	87	92

Girls attending school as a percentage of boys attending school: 1990/91 compared with 2001/02

4 Health and disease

Subject focus and objectives

Science
Knowledge and Understanding

Can you find relevant science information in different sources and give an opinion about the information justified by your knowledge and understanding of sciences?

Humanities
Skills (decision-making)

→ Can you make thoughtful decisions and relate them to real-world examples?

→ Can you formulate (build) arguments, make thoughtful judgements on events and draw basic conclusions and implications?

Organization and presentation

→ Can you communicate information that is relevant to the topic?

→ Can you present and express basic information and ideas in a clear and concise manner?

Area of interaction
Human Ingenuity

→ You are expected to reflect on the consequences of malaria, the history of innovation behind its treatment, and propose a potential course of action for its abatement in Africa.

Approaches to learning
Communication

→ being informed – including the use of a variety of media

→ informing others – including presentation skills, using a variety of media.

A mosquito sucks blood through its proboscis

Day 0, 23:00 Village near Siem Reap, Cambodia

As dusk fades into twilight, a female Anopheles mosquito flies over the rice paddy near the forest's edge. Only a few days before she emerged as an adult from the stagnant water. She heads toward the stilted huts of the village searching for a blood meal. She needs the protein-rich liquid to feed the eggs she is carrying.

Even in the darkness she knows there is a boy sleeping in the bed by the wall. Her impressive array of sensors picks up the CO_2 in his breath, the odour in his sweat and the heat from his body.

She lands undetected on his arm and plunges her proboscis (part cutting tool, part straw) through skin and fat into his bloodstream. Before she can begin to suck his blood she makes sure this puncture wound will not get clogged by injecting an anti-clotting agent contained in her saliva. Hidden in this spray is a deadly cargo of a dozen or so tiny malarial parasites which get into the boy's bloodstream. Within minutes they have taken up host in his liver and have begun to multiply.

Day 10, 07:00, A Stilted Hut in the Village

The boy has a hard time waking up and soon he starts to have waves of fever and chills. His body is wracked with pain. The parasites in his liver have multiplied many tens of thousands of times and are now on the loose in his circulatory system. They attack red blood cells and drain them of hemoglobin. They continue to multiply until they burst out of the cell to seek new victims.

His mother has seen these fevers rampage through the village, sometimes lasting several days. In this case however, things are more severe. Unless the boy gets some medical attention soon, his life is in danger.

Day 11, 18:30 Local Clinic

The boy's brain and other organs are now under attack. His breathing is difficult since so many blood cells have been destroyed. In trying to burn away the attackers, his fevers are literally cooking him from the inside. He has had several seizures and his back is now permanently arched, arms tensed, toes extended. A nurse administers something to help with the pain and it appears to bring some comfort. Without the correct medicine however, he is likely to become one of the almost one million children who die from malaria every year. He will add to the gruesome tally of one dead child every 30 seconds – four since you began reading this page, every hour, every day, with no end in sight.

vector • noun 2. An organism that transmits a particular disease or parasite from one animal or plant to another. 3. A chosen course or direction for motion

There are a host of good resources for finding out about malaria. You may wish to do your own research, or your teacher may provide you with some suggestions. After reading the above story and your other source(s), try to complete the structured response exercise on the next page.

1   Read the definition of a vector on page 45 and explain

   i what is the vector? _____

   ii what is being transmitted? _____

   iii to whom is it being transmitted? _____

   iv from who it is being transmitted? _____

2   Where does the parasite that causes malaria do its multiplying/reproduction?

   _____

   _____

3   Shallow or laboured breathing is a symptom at the end-point of the disease,
   however the parasite only attacks the circulatory system – specifically the blood.
   How are the two connected?

   _____

   _____

   _____

4   Mosquitoes are said to have poor eyesight, yet the mosquito in the story above was
   able to find the boy easily in the dark.

   i List the different ways mosquitoes are able to detect humans

   _____

   _____

   _____

   ii How could you design a "cloaking device" to keep yourself hidden from
   mosquitoes?

   _____

   _____

   _____

   iii DEET is the most effective mosquito repellent in the world – do some research
   and explain which of the detectors listed above it interferes with.

   _____

   _____

5 **i** Which property of blood requires mosquitoes to spray their saliva into the wound it has inflicted before filling up with a meal?

_____

ii How has the parasite taken advantage of this step?

_____

_____

_____

6 Create a 'lifecycle' for the disease from point of infection to end point. Be sure to make distinctions between parasite, vector, and host tissue. You may wish to include a diagram or diagrams.

7 Not everyone who contracts malaria dies of the disease. There are, in fact, four related parasites globally, each of which cause symptoms of varying degrees. Using the internet to research:

i Name the four parasites

_____

_____

ii Indicate which of the parasites are the most dangerous.

_____

_____

iii Indicate which parasites have begun to develop immunity

_____

_____

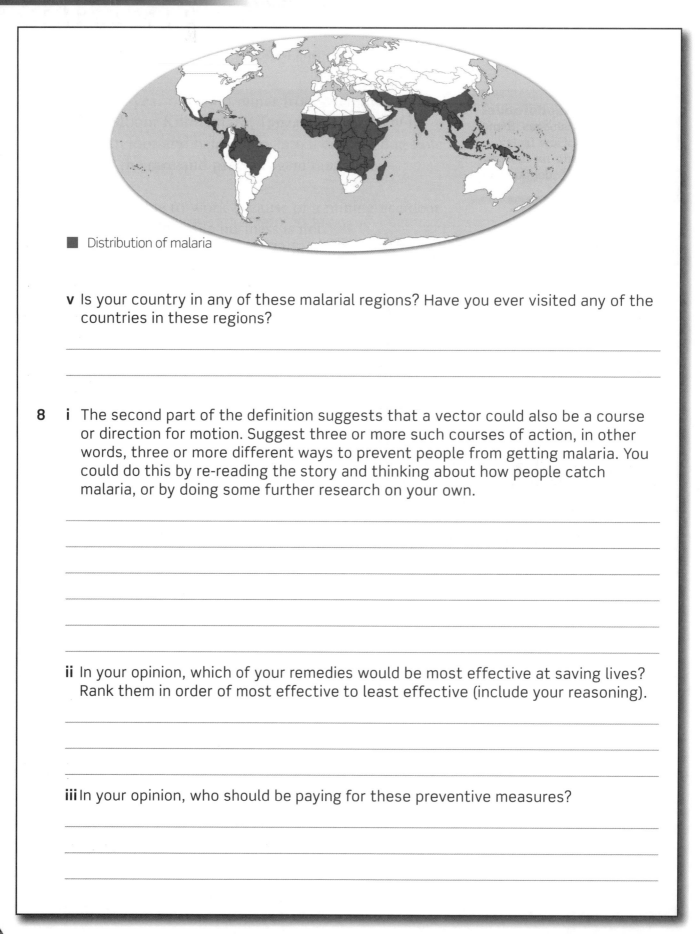

■ Distribution of malaria

v Is your country in any of these malarial regions? Have you ever visited any of the countries in these regions?

_____

_____

8 i The second part of the definition suggests that a vector could also be a course or direction for motion. Suggest three or more such courses of action, in other words, three or more different ways to prevent people from getting malaria. You could do this by re-reading the story and thinking about how people catch malaria, or by doing some further research on your own.

_____

_____

_____

_____

_____

ii In your opinion, which of your remedies would be most effective at saving lives? Rank them in order of most effective to least effective (include your reasoning).

_____

_____

iii In your opinion, who should be paying for these preventive measures?

_____

_____

_____

ISSUE

In other parts of this book, you may have had a chance to look at the impacts of global warming. Here's what the United Nations Development Report (2007/2008) has to say about climate change and malaria:

Rainfall, temperature and humidity are three variables that most influence transmission of malaria—and climate change will affect all three. Increased rain, even in short downpours, warmer temperatures and humidity create a 'perfect storm' for the spread of the *Plasmodium* parasite that causes malaria. Rising temperatures can extend the range and elevation of mosquito populations, as well as halving incubation periods.

For sub-Saharan Africa in particular, any extension of the malaria range would pose grave risks to public health. Some four in five people in the region already live in malarial areas. Future projections are uncertain, though there are concerns that the malarial range could expand in upland areas. More disconcerting still, the seasonal transmission period may also increase, effectively increasing average per capita exposure to malarial infection by 16–28 percent. Worldwide it is estimated that an additional 220–400 million people could be exposed to malaria.

A Heiroglyphic like this on the Temple of Denderah, Egypt reads "Do not get out the house after sunset during the weeks following the flooding"

You may wonder how some countries have been so successful at dealing with malaria while others have been less so. You may also wonder whether or not those countries without malaria should be doing something to help those that have it. Here is your chance to decide what you would do. The guiding question for the unit is:

How much would you spend to save a million children?

You will work with a group (assigned by your teacher) as a team of UN Health Workers dispatched to a conference in Malawi - a country where a staggering 100% of the population is at risk of malaria, to address the growing crisis in Sub-Saharan Africa.

In your group, make a list of what you think are the main issues in preventing the spread of malaria in the region.

You are likely to come up with many strategies. It might be helpful to break these into categories such as:

1 Preventing the breeding of mosquitoes

2 Preventing mosquitoes from visiting houses

3 Preventing mosquitoes from biting people

4 Treating people who have contracted malaria (this may include testing)

You may wish to address some of the following questions in your presentation:

A Which strategies are the most important/effective?

B How much is the plan likely to cost?

C How does this number compare with the average GDP lost in Africa due to malaria?

D How many lives are your recommendations likely to save?

E Who should pay for these measures?

F Why should anyone care about things that happen in other countries?

You should document how much of your research and ideas each member of the group contributed. You should also make sure there is an even split in the presentation duties, ensuring a voice for each member of the team.

DID YOU KNOW?

Chrysanthemums grown in Persia (modern-day Iran) and Dalmatia (modern-day Croatia) were valued for their insecticidal properties, the preparation of which was a closely guarded secret. Eventually the active ingredients, known as pyrethrins, were isolated and are used today to treat mosquito nets.

DID YOU KNOW?

The Chinese described the qinghao plant as early as the second century BCE as a remedy for fevers. The active ingredient artemisinin is today's most powerful anti-malarial drug when used in combination.

Our project plan

You may wish to address some of the following questions in your presentation:

1   Which strategies are the most important/effective?

2   How much is the plan likely to cost?

3   How does this number compare with the average GDP lost in Africa due to malaria?

4   How many lives are your recommendations likely to save?

5   Who should pay for these measures?

6   Why should anyone care about things that happen in other countries?

Project evaluation

What I enjoyed about our project and why?

_____
_____
_____
_____
_____
_____
_____

What aspect didn't I enjoy and why?

_____
_____
_____
_____
_____
_____
_____
_____

What did I/our team do really well?

_____
_____
_____
_____
_____
_____
_____

What did I/our team do less well?

_____

_____

_____

_____

_____

_____

How well did we work together as a team? (If applicable)

_____

_____

_____

_____

_____

_____

_____

What would I/we do differently next time?

_____

_____

_____

_____

_____

_____

Which group's project work did I like best and why?

_____

_____

_____

_____

_____

_____

_____

Some scientists believe that as many as half of all people who have ever lived have had malaria. There is no doubt that malaria has killed more people since the beginning of recorded history than any other disease. No culture, civilization or empire has been untouched by malaria, including the Egyptians, Greeks, Romans, and Incas. Many of today's most powerful preventive measures and treatments for malaria have their basis in ordinary plants. Quinine is one such treatment and its discovery, development and distribution has a long and varied history. Create a timeline for the antimalarial treatment quinine. A checklist of some of the more important events are given below and you may add as many details as you like.

Major events in the development and history of quinine

1 Incan pre-history in regards to the cinchona tree.

2 Introduction of malaria to the New World (South America)

3 "Jesuit bark"

4 Isolation of active ingredient from bark

5 Cultivation of cinchona trees outside of Peru

6 Axis occupation of Java, crisis for allies

7 Use of quinine in colonization of Africa and India by British

8 Use of quinine in Suez canal

9 Rise of resistance to quinine

10 Identification of malarial parasite and mosquito as malarial vector.

A botanical print of the cinchona tree. Quinine comes from its bark.

You should do your best to put a date (time-tag) on each event and sort these in order of when they happened (chronologically). Each event should contain a brief but detailed explanation. A list of sources for your information should be included. Any photos/diagrams/figures that you include, like the one above, should also be referenced.

Global trade and development

Unit question
How hard should children work?

Children at a brick factory in Bangladesh

Subject focus and objectives

Technology
Can you work through the stages of the design cycle?

Investigate:
➔ Can you ask useful questions about a design brief and research and organize the responses?
➔ Can you list the requirements for the product and do you understand what a design specification is?

Plan:
➔ Can you create a design, compare it with the design specification, and make a realistic plan of how to create the product?

Create:
➔ Can you follow a plan, monitoring progress and adapting the plan if necessary as you go along?
➔ Can you produce a quality product which matches the design brief?

Evaluate:
➔ Can you evaluate the success of your product through research and reflection?

Arts
Knowledge and understanding
Can you use some basic language and have a simple understanding of some of the concepts and processes that support your current work?

Application
Have you learned skills and developed techniques and used these ideas to create and present art, with your teacher's guidance?

Area of interaction
Community and service
You are expected to show willingness and the skills to respond to the needs of others, coming up with solutions to actively resolve issues within communities.

Approaches to learning
Thinking
Planning, including storyboarding and outlining a plan.

Wilson is a 12-year-old child miner from Mererani, in the foothills of Mount Kilimanjaro, Tanzania. Every day he descends, barefoot and helmetless, into a crudely dug hole to search for the rare and precious gem tanzanite.

His father is unable to work because of a mining accident and his mother's vegetable business is not enough to keep the family going. Although he frequently works twelve-hour days he sometimes comes home empty handed. There are no guaranteed wages here. If he finds no tanzanite he is not paid a penny.

His lungs ache from the dust he must breathe in and he lives in constant fear of dynamite blasts, collapsed tunnels, and falling rocks.

Once home, he must go out and gather water for the cooking and cleaning. After he has helped his mother with chores he finally gets to eat before going to bed. He has to get an early start – his life, and that of his family, depends upon it.

labor/labour • noun 1 work, especially hard physical work. • verb 2 work at an unskilled manual job.

Twelve-year-old Alejandra wakes before dawn and prepares for another day of collecting *curiles* (small molluscs) in the mangrove swamps off the island of Espiritu Santo in El Salvador.

To keep the swarms of mosquitoes at bay she smokes more than a dozen cigars a day. These will cost her not only in terms of her wages, but also her health. She often spends more than 14 hours in the swamp, digging barefoot in the mud, sometimes in the dark, as she dodges the ever changing tide.

Struggling to keep awake, she takes pills to prevent her from falling asleep. She is extremely thin since she has little time to eat, and food would only take more of the $1.40US she makes if she is fortunate enough to collect two full baskets. She does not go to school and has no time for friends. She is busy helping support her seven brothers and sisters.

Sukumar's father borrowed 1700 rupees ($35US) from a money lender to fix the roof of their house. In order to guarantee the loan, Sukumar was given to the lender as a bond until the debt was paid.

He was forced to work making *bidis* (hand-rolled cigarettes made from tobacco wrapped in tendu leaves) at the owner's factory in Gudiyattam, a town in the Tamil Nadu province of southern India. He was forced to sit cross-legged in a cramped room, without a fan, for 15 hours a day in order to meet his daily quota of 4000 bidis. He received no food and was beaten if the quota was missed or if he asked to go to the toilet too often. When he complained about mistreatment, his father told him he had to do whatever the owner demanded.

Questions

Reflection

1   What were the reasons for the children working?

_____

_____

2   What else did their stories have in common? You should be able to list at least three. Use complete sentences in your answer.

_____

_____

3   Which child do you think had the worst job? Explain why?

_____

_____

4   What either surprised or upset you when you read these stories?

_____

_____

_____

Sukumar, Alejandra, and Wilson may live oceans apart, but they are connected in their misfortune with millions of other children around the world who are forced into child labor. Believe it or not, their stories are not the worst case. Many children are trafficked (sold and sent away from their families to foreign countries), forced into armed conflict (child soldiers), crime, slavery, and other dangerous, damaging and degrading activities.

In trying to address the problem of child labor, it must first be defined. One of the simplest definitions of child labor is - any work that *deprives a child of their childhood.*

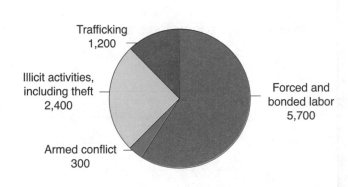

Children in worst forms of child labor in thousands
(International Labour Organization, 2000)

Although you most certainly don't think of yourself as a child, you are likely of a similar age to the people in the stories above. To compare your life with theirs, try to make a list of the activities you do in an average week. These should include things you do at school, at home, and outside the home, and should include structured and leisure time. It may include examples of labor. To add some meaning to this exercise, try to estimate a time spent on each activity. You could use a table similar to the one below:

Activity	Hours/Day	Hours/Week
Watching television	4	28
Internet (school)	1	5*
Internet (social/leisure)	2	14
Football	varies	4
School	5	25

* No schoolwork on weekends.

When you have finished, compare your table with one or two other members in your class and answer the following questions:

1 Which activities did you have in common?

_____

_____

2 Which activities were different?

_____

_____

3 How were the numbers of hours spent on activities similar?

_____

_____

4 How were the numbers of hours spent on activities different?

_____

_____

5 How many hours of your week would you consider yourself to be doing "work".

_____

_____

Once you have completed your comparison, try to repeat the exercise for one of the people in the stories above. This will likely not take you very long. When you compare your chart with the children above, ask yourself the following two questions:

1 Who works harder, the child in the story or me? Explain your answer.

_____

_____

_____

2 How are either of us being deprived of a childhood?

_____

_____

_____

_____

Trying to define what a 'child' is, and what 'labor' means can be a tricky business.

It is important to note that not all work done by young people is necessarily a bad thing. Surely the girl selling oranges on the streets of Dakar in order to help pay her school fees, or the boy delivering newspapers in Calgary for pocket money aren't being deprived?

For the purpose of our discussion child labor includes:

→ work performed by children under the age of 15
→ long hours of work on a regular or full-time basis
→ abusive treatment by the employer
→ no access, or poor access, to education

The most extreme forms of this labor include work that is mentally, physically, socially or morally dangerous and harmful to children.

Before doing the tasks on the next few pages, you may wish to do a bit more research on child labor. It is a very difficult thing to find facts about. Statistics are highly variable since most countries have agreements in place banning the practice. Your teacher may have some material to share with you, or you could visit any number of websites dealing with the issue. These include the International Labour Organization (ILO), the United Nations Development Program, Save the Children, Unicef, Oxfam, and many others.

ISSUE

Education is the theme of Unit 3 in this book and it is considered the number one remedy for the problem of child labor. In fact, the ILO estimates that although it would cost 55 billion US dollars a year (from 2001 to 2020) to eliminate child labor and enroll children in school, the economic benefits would amount to more than $5 trillion US over the same period. This means that the benefit of sending a child to school outweighs the cost of lost labor by almost 7 to 1!

In case you were wondering where you could raise $55 billion US a year, consider that we spend 10 times as much on wars and military expenditures, and 20 times this amount every year paying the interest on our governments' debts.

Activity

You have been hired by a new organization called *Work to School.** It is dedicated to eradicating child labor and getting children in school. Part of your job is to create a radio, television, or internet announcement for the organization (**Part I**). You will also be helping them to create a logo for their website (**Part II**).

* Your first order of business might be to change the name of the organization if you can think of something more suitable or engaging.

Our project plan

Part I

To help you create your broadcast you will be using the information design cycle here.

Evaluate the use of the design cycle:

Identify the problem

You have been asked to design a broadcast for the purpose of raising awareness of the issue of child labor and to encourage people to get more information by visiting the website (you can choose an appropriate imaginary web address too). This broadcast is your **product**.

A **design brief** is a means of generating questions that will guide the investigation, planning, creation, and evaluation of your product. Here are some possible research questions that could get you started on your design brief:

1 Will I produce a radio or television or internet-based broadcast?

2 What equipment will I need to produce my broadcast?

3 Can I make this broadcast alone or will I need any help (technical help, supporting characters/voices, etc.)?

4 How long should my broadcast be?

5 Which information should I include in my broadcast?

6 How is my broadcast going to make a difference to the problem of child labour?

7 Who is my target audience?

8 How will I test whether or not my broadcast was effective?

9 What does a good broadcast look or sound like?

Feel free to add any other questions to this list.

Once the above questions are answered you will have a working template. This is called a **design specification** or **design spec** for short. You can use the template to create possible designs and to help you decide which of your designs is 'the best'. The answers to the above questions must be kept in something called a design folder.

Design

Once you have completed your design spec, you will need to come up with **at least two** possible designs for a broadcast. Each of your potential designs needs to follow the recipe laid out in your design spec. For example, if you chose a 30-second radio spot, you will need to design two different radio spots, not one radio and one television.

If you are having trouble coming up with different versions of the same product you should consider that there is more than one way to get your message across. This could involve the content (emotional, funny, serious, factual, etc.) or the presentation (flashy, colourful, loud, toned-down, stylized, etc.), or the technique (animation, voice-overs, actors, music, images, sound/visual effects).

All of your ideas/scripts/storyboards need to go into your design folder. Once completed you must choose the best design. Often, students will get an idea that they like and spend most of their time on the 'real' version, and add a token second and third design. These students are often less successful. The more high quality designs you have to choose from, the better your final product will be.

You must **justify** your choice of design. This will have a lot to do with how closely the design meets the criteria laid out in your design spec. If you have a struggle choosing the best design, you know you have done a good job.

Plan

Now that you know what product you are working towards you can plan. You will need to make

⇒ a list of all materials required.

⇒ a timeline for how and when you will complete certain portions of your task.

If this timeline is annotated (includes specific information on the plan) you may not need to write up a separate recipe or instruction sheet for how you will complete your design. Your teacher will assign an appropriate time length for this project which you will need to factor in to your timeline.

There is almost no chance that your original plan and your final product will be exactly the same. Any changes you make along the way and their justification will go in the following section.

What?

Every time you do something new – take a photo, add some text, do some research – you should update your folder.

Where?

It would be helpful if your design folder used the headings you see here (Investigate, Design, etc.). This will help you organize your work and help your teacher assess your product.

Why?

By staying organized, you avoid having to try to remember when you performed which step. A bit of work now will save you lots of time later.

Create

Once you have thought about the problem, chosen the best design, and made a plan to create your product, you must actually create your broadcast. While you are creating your product you need to document the process. This could include photos, videos and diary, journal or log entries. As was mentioned above, staying organized is crucial at this stage and all documentation should carry a time/date stamp.

Evaluate

Once you have completed your broadcast you will likely be anxious to show off your hard work. Remember that you were asked to think about ways of testing how effective your product was. If this were an actual broadcast being aired on television or radio, you would know fairly quickly by the change in the number of 'hits' on your website, how successful your campaign has been.

A survey is a great way to collect evidence and evaluate a product. Getting people to take them is another issue, but luckily your classmates and peers are all in the same boat and are likely to help you.

There are many ways to create surveys, but freeform answer questions like "what did you like about my broadcast" might not elicit the kinds of responses you are looking for. You might try a rating system. For instance

Q5 The voiceover in the broadcast was clear and effective.

☐agree ☐agree somewhat ☐disagree

You should also make a list of all the things you might do differently or that could be improved if you were to do this activity again.

Part II

You have also been asked to help design the new logo for the organization *Work to School*. Before you can create your own logo, you should do some thinking about what makes a good logo. This will include collecting logos, classifying them, reflecting on them, and finally creating your own.

1 Collection

⇨ Begin collecting logos from various media including print and electronic.

⇨ This may be an ongoing process over a few classes or an assigned task for homework.

For the purpose of this exercise, these should be limited to not-for profit organizations. Avoid using company logos.

⇨ You should keep these in a file for future reference.

⇨ Make a note of the source of where you collected the logo. It may be interesting later on to consider which types of publications and websites feature which types of organizations.

2 Analysis

⇨ Once your collection is complete, separate your logos into two or three piles. You could call these: "Like a lot", "OK", and "Don't like".

⇨ For the first group of logos, ask yourself if they have certain characteristics or qualities that caused you to place them in the "like a lot" pile?

⇨ Repeat the exercise for the other two piles.

⇨ Once complete you should have a fairly long list of things you would consider when designing your own logo.

Here are some prompts to help you with your list:

⇨ colourful/limited palette/ black and white?

⇨ simple/complex?

⇨ realistic/graphic/ cartoonish?

⇨ symbolic?

⇨ attention-grabbing/subtle?

⇨ includes organization name/organization abbreviation/doesn't include name

3 Design

Now that you have considered which characteristics of logos you like best, you should set about creating the *Work to School* logo. Your teacher will discuss with you which media you should use to create your logo and appropriate dimensions. Although this is not a tech project, you may wish to generate several different designs and choose the *best* to produce formally. All designs should go in your developmental workbook.

logo • noun (pl. logos) an emblematic design adopted by an organization to identify its products

Moving on

Some good news...

Each of the three children mentioned on pages 56 and 57 have been removed from their work environments and been given the chance to go to school.

All three have grasped that chance and made the best of it. In fact, Sukumar enrolled in an engineering program at university. With the money he makes from his new job he plans to pay for his sister to go to school.

Craig Kielburger was only 12 years old when he read a newspaper article about a Pakistani boy who was killed because of his attempts to give child laborers a voice. The article had a profound effect on him and set in motion a series of events that led him to create *Free the Children*, an organization that has grown into the world's leading youth-driven charity reaching more than one million people. They have built over 500 schools worldwide providing daily education to more than 50,000 children, helping to break the grip of child labor. Craig proves that one person can make a difference.

Craig Kielburger

Extension

What other examples can you find of young people who have made a difference in the lives of other young people especially in the area of child labor?

_____

_____

_____

_____

_____

_____

_____

6 Environmental sustainability

Unit question
How big is the mess our parents are leaving us?

Subject focus and objectives

Could this happen?

Science
One world
- → Can you make comments on the ways in which science is applied and used to solve local and global problems?
- → Can you show understanding that science is part of the world by giving examples and commenting on ways in which science affects life, society and the world?

Attitudes in science
- → Can you carry out scientific investigations, with guidance, using materials and techniques safely and skilfully?

Mathematics
Investigating patterns
- → Can you recognize and describe simple patterns in words/diagrams?
- → Can you arrive at a result or set of results and make predictions based on extending the pattern(s)?

Reflection
Can you consider the reasonableness of your findings and importance of your results?

Area of interaction
Environments
You are expected to develop an awareness of the effects of our actions and inaction on the environment and discuss your responsibilities.

Approaches to learning
Transfer
Inquiring in different context - including changing the context of an inquiry to gain various perspectives.

Starting points

In 1958 a young Caltech scientist named Charles David Keeling began collecting data on atmospheric CO_2. He did this at the South Pole and on a remote dormant volcano, Mauna Loa, in the middle of the Pacific Ocean.

perspective • noun A particular way of regarding something

Why in the world?
If you were trying to make accurate measurements of CO_2 in the atmosphere, why would you choose such remote locations?

CO_2 is a very common gas produced by all animals and plants when they respire, but it makes up only a small percentage of the mixture we call air. It is so small, in fact, that scientists measure it in parts per million (ppm).

Carbon dioxide has remained in relative balance for hundreds of thousands of years. It is one part of a dynamic and complex system called the Carbon Cycle.

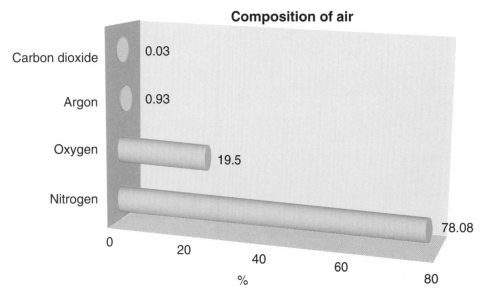

Composition of air

	%
Carbon dioxide	0.03
Argon	0.93
Oxygen	19.5
Nitrogen	78.08

Historic data on CO_2 levels are gathered from ice core samples at the Antarctic. As you can see, over the past 400,000 years, the concentration has fluctuated roughly between 200 and 300 ppm.

industrial revolution • the growth of industry by the use of machines in the 18th and 19th centuries

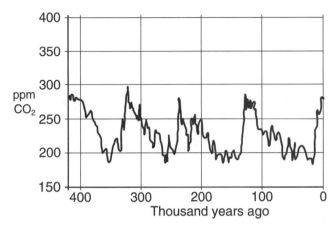

Since humans began mining, refining and burning fossil fuels to support the energy requirements of the industrial revolution, trillions of tons of CO_2 have made their way into the atmosphere.

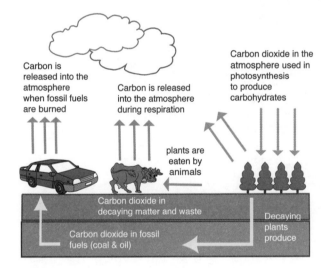

Extension

You could learn more about the Carbon Cycle by doing some research on your own. Your teacher might be able to provide you with information as well. Some questions you could think about as you do your research include:

1   How much is a gigaton?

2   How does deforestation affect the cycle?

3   How does reforestation affect the cycle?

4   Are there any alternatives to fossil fuels that might reduce inputs of CO_2 to the atmosphere?

Deforestation in rainforest

Starting points

Recognizing patterns

Here is data from 1958 to 2007 on atmospheric CO_2.

Year	CO_2 (ppm)	Year	CO_2 (ppm)	Year	CO_2 (ppm)	Year	CO_2 (ppm)
1958	315.64	1971	326.92	1984	345.27	1997	364.89
1959	316.56	1972	328.53	1985	346.52	1998	367.62
1960	317.16	1973	329.63	1986	347.82	1999	368.59
1961	317.98	1974	330.5	1987	349.9	2000	370.33
1962	318.67	1975	331.56	1988	352.16	2001	271.83
1963	319.33	1976	332.75	1989	353.56	2002	374.45
1964	319.68	1977	334.55	1990	355.15	2003	376.71
1965	320.34	1978	335.86	1991	355.91	2004	378.25
1966	321.97	1979	337.78	1992	356.27	2005	380.81
1967	322.85	1980	339.3	1993	357.59	2006	382.59
1968	323.73	1981	340.91	1994	359.65	2007	384.64
1969	324.97	1982	341.61	1995	361.29		
1970	325.92	1983	343.79	1996	362.78		

Use the data to make a scatter plot on the axes provided.

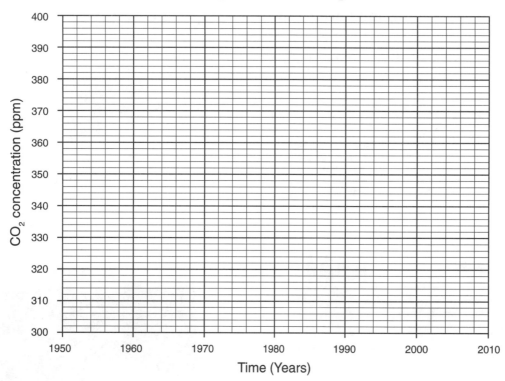

If you have access to Excel or other graphing software your
teacher may provide you with a file of the data to use.

1   What happens to the concentration of carbon dioxide over the time?

   _____

   _____

2   Would you say this is a straight line or a curved line?

   _____

   _____

The graph here gives you an opportunity to practice your skills at predicting into the future. This is the very same kind of mathematical modelling that climate scientists are practicing right now.

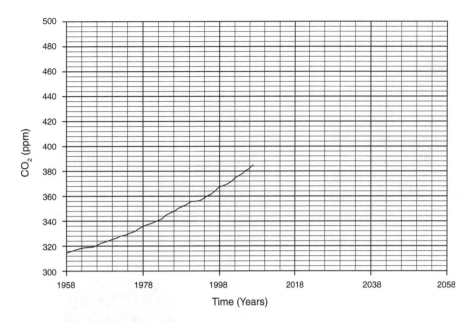

Extending the pattern

If global life expectancy trends continue, the majority of you will still be around in 2058 to mark the 100-year anniversary of data collected at Mauna Loa. Using a 'best line' extrapolate the data and make a prediction about the amount of CO_2 in the atmosphere at this time.

Report this value here. _____

Quantify

1   Can you formalize the link between CO_2 and time by quantifying it? For instance; "A change of *X* years results in a change of *Y* ppm of CO_2."

You could start by finding out how much the CO_2 concentration increases over a period (such as 20 years).

Change in CO_2 concentration (ppm)/number of years

= average increase per year (ppm per year)

_____

_____

_____

_____

2   Use this mathematical model to predict the concentration of carbon dioxide in 2058.

_____

_____

_____

Why should we care about CO_2?

Try the experiment below

Problem:

What is the effect of increasing the number of layers of cling wrap on the temperature inside a cup?

Hypothesis:

In the space below, make a prediction on what will happen. Be sure to include an explanation if possible.

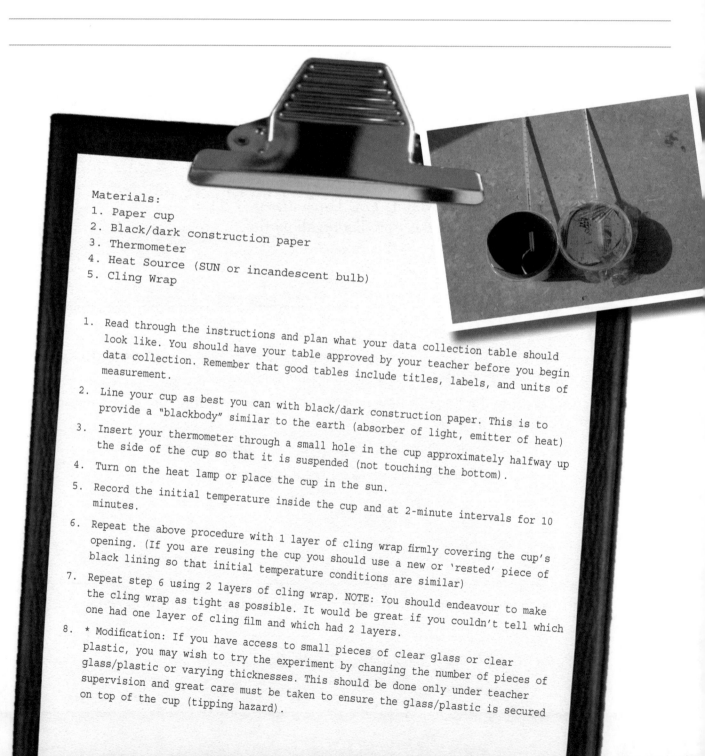

Materials:
1. Paper cup
2. Black/dark construction paper
3. Thermometer
4. Heat Source (SUN or incandescent bulb)
5. Cling Wrap

1. Read through the instructions and plan what your data collection table should look like. You should have your table approved by your teacher before you begin data collection. Remember that good tables include titles, labels, and units of measurement.

2. Line your cup as best you can with black/dark construction paper. This is to provide a "blackbody" similar to the earth (absorber of light, emitter of heat)

3. Insert your thermometer through a small hole in the cup approximately halfway up the side of the cup so that it is suspended (not touching the bottom).

4. Turn on the heat lamp or place the cup in the sun.

5. Record the initial temperature inside the cup and at 2-minute intervals for 10 minutes.

6. Repeat the above procedure with 1 layer of cling wrap firmly covering the cup's opening. (If you are reusing the cup you should use a new or 'rested' piece of black lining so that initial temperature conditions are similar)

7. Repeat step 6 using 2 layers of cling wrap. NOTE: You should endeavour to make the cling wrap as tight as possible. It would be great if you couldn't tell which one had one layer of cling film and which had 2 layers.

8. * Modification: If you have access to small pieces of clear glass or clear plastic, you may wish to try the experiment by changing the number of pieces of glass/plastic or varying thicknesses. This should be done only under teacher supervision and great care must be taken to ensure the glass/plastic is secured on top of the cup (tipping hazard).

Results

You may wish to plot line graphs for the change in temperature of each cup over time. You may wish to plot the final temperature in each cup against the number of layers of cling wrap. Discuss which option would best represent your data with your teacher.

Discussion

What trend(s), if any, did you notice in your data?

_____

_____

_____

What can you conclude about the effect of layers of cling wrap on the internal temperature of the cup?

_____

_____

_____

Which errors, if any, were present in this investigation?

_____

_____

_____

What kinds of improvements could you make to avoid these limitations?

_____

_____

_____

What type of further research could this lead to?

_____

_____

_____

Transfer

In some ways, your cling wrap acts like CO_2 in our atmosphere, allowing light energy to enter but trapping some re-emitted heat. If your graphs from earlier have shown increasing amounts of CO_2 in the atmosphere, what would you now predict about the temperature on Earth?

_____

_____

_____

Several other greenhouse gases contribute to the heating trends you have observed. How is it that some gases are able to hold heat in the atmosphere while others – notably the two most common gases, nitrogen and oxygen, are not? The answer lies, in part, with the number of atoms. If gases have three atoms (are triatiomic) or more, they have the ability to absorb and re-radiate heat.

Anthropogenic • adjective (chiefly of pollution) originating in human activity

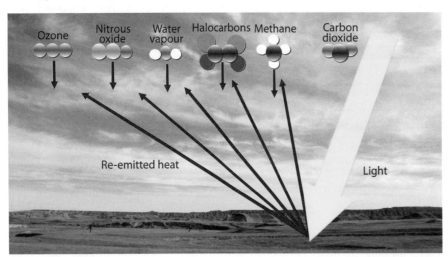

From the horse's mouth?
This idiom means from the original source of information. If you found some information about one of these gases on a website, for example, annual global methane emissions, how would you quote it? Should you credit that website as your source, or would it be better to find out where the website got the information and visit that website?

Research

You could do some research on one or all of the greenhouse gases shown in the diagram in a format similar to the table below. You would, of course, need to make a list of where you got your information from.

Table for greenhouse gas research

Gas	Source		Greenhouse rank	Emission rank
	Natural	Anthropogenic		
Halocarbons				
Methane				
Nitrous oxide				
Ozone				
Water vapour				

Be sure to distinguish between natural and human sources of gases. If your teacher assigns individual gases to groups then you will need to discuss with the class how to determine ranks. In the case of greenhouse rank, you will need to make a value judgement, as there may be no hard data. The greenhouse rank has to do with how 'good' the gas is at capturing and re-emitting heat, and how long it tends to stay in the atmosphere. Emission rank has to do with how much of the gas is produced worldwide and how long it's been accumulating.

Our project plan

You may have heard about dissenting opinions regarding

⇒ climate change

⇒ science and scientists making predictions

⇒ the economic and political trade-offs making changes would require.

With so many different groups of people having a stake in the cause, impact, and prevention of climate change, you may wonder who you can rely on for information. How would you feel about a report like the one advertised here?

Just such a report is available from the International Panel on Climate Change. You could have a look at some of the predictions these scientists are making about the impacts of climate change in your lifetime. A summary of these is given here (the numbers refer to the relevant section of the main report).

> 2500+ SCIENTIFIC EXPERT REVIEWERS
>
> 800+ CONTRIBUTING AUTHORS AND
>
> 450+ LEAD AUTHORS FROM
>
> 130+ COUNTRIES
>
> 6 YEARS WORK
>
> 1 REPORT

For the next two decades, a warming of about 0.2°C per decade is projected for a range of SRES emission scenarios. Even if the concentrations of all greenhouse gases and aerosols had been kept constant at year 2000 levels, a further warming of about 0.1°C per decade would be expected. {10.3,10.7}

Continued greenhouse gas emissions at or above current rates would cause further warming and induce many changes in the global climate system during the 21st century that would very likely be larger than those observed during the 20th century.{10.3}

Anthropogenic warming and sea level rise would continue for centuries due to the time scales associated with climate processes and feedbacks, even if greenhouse gas concentrations were to be stabilised. {10.4, 10.5, 10.7}

In other units we examine further causes of climate change, and especially their potential impacts.

Reflection

Now that you have researched various aspects of climate change, you can start to answer the unit question:

How big is the mess our parents are leaving us?

Other things to consider in your reflection:

→ How close is your temperature prediction for mid-century (2050) to their prediction?

→ Are the people making decisions for you truly representing your best interests?

→ How are your interests being represented?

→ How can you have a voice in these decisions?

→ What do you think you could do about climate change?

Plan here how you will present your answer and its justification, drawing on your work earlier in the unit:

Project evaluation

What did I enjoy about this unit, and why?

_____

_____

_____

What aspect didn't I enjoy and why?

_____

_____

_____

What did I/what did our team do really well?

_____

_____

_____

What did I/what did our team need to improve?

_____

_____

_____

What would I do differently next time?

_____

_____

_____

What challenges did I/our team face while working on this unit?

_____

_____

_____

What do I know now, that I didn't know before working on this unit?

_____

_____

_____

Why does this topic matter?

_____

_____

_____

Moving on

In case you think one person can't make a difference, you should consider the case of Severn Suzuki. In 1992, at 12 years of age, she and a few friends raised enough money to fly from their home in Vancouver, Canada, to the Earth Summit in Rio de Janeiro. While there, she delivered a powerful and moving speech to world leaders addressing the very notion investigated in this unit - how leaving a mess for future generations to clean up is not acceptable.

Here is an **extract from the speech:**

Here, you may be delegates of your governments, business people, organizers, reporters or poiticians - but really you are mothers and fathers, brothers and sister, aunts and uncles - and all of you are somebody's child.

I'm only a child yet I know we are all part of a family, five billion strong, in fact, 30 million species strong and we all share the same air, water and soil -- borders and governments will never change that.

I'm only a child yet I know we are all in this together and should act as one single world towards one single goal.

In my anger, I am not blind, and in my fear, I am not afraid to tell the world how I feel.

In my country, we make so much waste, we buy and throw away, buy and throw away, and yet northern countries will not share with the needy. Even when we have more than enough, we are afraid to lose some of our wealth, afraid to share.

This speech is widely available on the internet.

Do take 5 minutes to watch it if you get the chance. Perhaps you could encourage your parent or guardian to watch it with you. It is possible they haven't thought about their part in climate change since they were your age.

Acknowledgments

The Publisher and authors would like to thank the following for permission to use photographs and other copyright material:

Cover photo Phi2/iStockphoto; Corel; p7 Tuca Vieira; p8 Andrey Prokhorov/iStockphoto; p9t Oxfam; p9b Google; p13 Dave Long/iStockphoto; p15 Sean Warren/iStockphoto; p18 Edward Parker/Alamy; p19 AFP/Getty Images; p21t Photodisc/Oxford University Press; p21b Ian Shaw/Oxford University Press; p22l Li Erben/Kipa/Corbis UK Ltd.; p22r Omar Torres/AFP/ Getty Images; p22cl Condè Nast/Corbis UK Ltd.; p22cr Brad Edelman/Corbis UK Ltd.; p22b Tim Mosenfelder/Corbis; p23 Photodisc/Oxford University Press; p24t Vasko Miokovic/ iStockphoto; p24b Adalberto Roque/AFP/Getty Images; p25 The London Art Archive/Alamy; p26 Aaron Asterley/Alamy; p30l oneclearvision/iStockphoto; p30 Claudia Dewald/ iStockphoto; p31 Jerome Delay/AP/PA Photos; p42 fotoshoot/Alamy; p43 Christopher Badzioch/iStockphoto; p44 Nic Cleave Photography/Alamy; p49 Tawfik Deifalla/iStockphoto; p50t Thomas Mounsey/iStockphoto; p50b Scott Bauer/US Department Of Agriculture/ Science Photo Library; p54 Gianni Dagli Orti/Corbis UK Ltd.; p55 G.M.B. Akash/Panos Pictures; p56 IRIN news; p57t Antonio Rosa/www.ccoo.cat/noesunjoc/Comissions Obreres de Catalunya; p57b Sophie Elbaz/Sygma/Corbis UK Ltd.; p59t Oxford University Press; p59b image100/Oxford University Press; p66 Tom Hanson/AP/PA Photos; p67 John Blackford/AA Reps, Inc.; p69 Joseph Luoman/iStockphoto; p70 Barclay Lelievre; p74 Greg Okimi/ iStockphoto.